THE EXPRESSIVE LEADER

How To Deliver Your Message Effectively, Confidently, And Have The Impact You Want On The Audience

Jonathan Li

The Expressive Leader: How To Deliver Your Message Effectively, Confidently, And Have The Impact You Want On The Audience

Published by CreateSpace Independent Publishing Platform

Cover design by Pixel Studio

CONTENTS

INTRODUCTION

I'm passionate about helping business leaders deliver their message effectively and confidently and have the impact they want on their audience. I've interviewed successful leaders to write this book.

If you want to become the best, you must learn from the best.

I understand you're busy. Who has time for a 300-page book? This powerful book gets to the point: you'll discover the secrets to become an effective and confident presenter... in minutes.

Carry this book with you. When you need to deliver a message, read it. This quick read will help you make a big impact in front of the audience.

I hope you enjoy this book. It will help improve your presentation skills and grow your business.

PART I:

FIND YOUR MESSAGE

1.
How To Find Your Message In Ten Minutes

"We all have a life story and a message that can inspire others to live a better life or run a better business. Why not use that story and message to serve others and grow a real business doing it?"

- BRENDON BURCHARD

Are you unsure what your message is? Don't know what message you need to deliver?

Here's the solution: Find your message.

To find your message in ten minutes, ask yourself these two questions:

1. WHAT DO I WANT THE AUDIENCE TO THINK, FEEL, OR DO *DIFFERENTLY* AFTER MY PRESENTATION?

For my TEDx talk, *How To Speak with Confidence,* I want the audience to think that becoming a confident speaker can be easy.

Invest five minutes to think about your next presentation. How do you want to improve people's lives? It can be a new thought, feeling, or action step.

2. IF I COULD HAVE ONE SENTENCE TO SHARE WITH PEOPLE AND THEY WOULD LISTEN, WHAT WOULD I SAY?

Create your message in one sentence, so the audience will remember what you say and take action.

The message of my TEDx talk is, "To speak with confidence: prepare, practice, perform." Short and memorable.

Take 5 minutes to write down your message in one sentence.

By answering these two questions, you will have a clear message that changes people's lives. You will know what to say in presentation and have a positive impact on the audience.

2.
Ten Examples Of Memorable Messages To Inspire You

"What makes things memorable is that they are meaningful, significant, colourful."
- JOSHUA FOER

Creating a memorable message can be challenging...

That's why I've created ten memorable messages to inspire you.

Feel free to model them:

- To speak with confidence: prepare, practice, perform.

- To become the best, you must learn from the best.

- Practice doesn't make perfect; practice with feedback makes great improvement.

- To gain confidence, give up your ego.

- If you can't even motivate yourself, how can you motivate others to take action?

- To change the world, change yourself first.

- Great leaders focus on the audience; poor leaders focus on themselves.

- The purpose of presentation is to inspire the audience to take action.

- Either you control fear, or it will control you.

- Your confidence comes from what you do before speaking.

PART II:

DELIVER YOUR MESSAGE EFFECTIVELY

3.
Three Mistakes Most Business Leaders Make When Delivering Their Message (And How To Avoid Them)

"The successful man will profit from his mistakes and try again in a different way."
- DALE CARNEGIE

I hate my presentation...

"Jonathan, I don't get your message... Can you say it again?"

How can people misunderstand my message? I've prepared this presentation for months!

Has this situation ever happened to you?

Most business leaders make people

misunderstand their message because they make these three presentation mistakes.

Avoid these stupid mistakes so people will get your message and change their lives for the better.

MISTAKE 1: THEY HAVE MORE THAN ONE MESSAGE

Business leaders want to be productive, so they try to include ten messages in a 20-minute presentation. Result? No one remembers what they say. They are wasting time.

TIP: FOCUS ON ONE MESSAGE

What do you want the audience to think, feel or do *differently* after your presentation? Focus on that one important message. Your audience will understand what you say and change.

MISTAKE 2: THEIR MESSAGE IS NOT MEMORABLE

"You should buy my products because

they are the best in the world. They are convenient, trendy, awesome, fun, and help you get results fast."

Seriously? Who will remember this? A confused mind don't take action.

TIP: SUMMARIZE YOUR MESSAGE IN ONE SENTENCE

"Getting online coaching will save you years of time and frustration."

This message is punchy and includes the big benefits.

A clear message inspires action. Keep it short.

MISTAKE 3: THEY HAVE NO STORIES TO SUPPORT THEIR MESSAGE

Facts and data are boring. Stories are interesting.

People are emotional. Stories help you engage the audience's emotions such as hope, love, fear, anger, and joy.

When you engage the audience with emotions, they will resonate with what you say.

TIP: DELIVER YOUR MESSAGE WITH STORIES

Powerful stories are around you; you just need to find them.

Here are three types of stories that deliver your message effectively:

a) *Your Personal Experiences.* What experience has changed your life? It could be "the biggest mistake I've had in my career", "the first job I had" or "my first sales experience." Anything relates to mistakes you've made and your first-time work. These stories make you human.

b) *Client Case Studies.* Think about the results you've helped clients get. To prove my ability to help business leaders succeed, I can share how I helped a nervous business leader deliver the message effectively and confidently and have the impact he wanted on the audience.

c) *Celebrity Stories.* Who doesn't like stories from famous people? Movies, artists, politicians, experts... Do you know the world's greatest life coach has a coach too? After speaking at too many events, Tony Robbins lost his voice... almost. Miserable, he worked with a voice coach, Roger Love. With effective voice exercises, Roger saved Tony's voice and career. Tony now has a powerful voice and inspires millions of people.

Choose one type of stories for your next presentation. Deliver your message effectively and change lives.

4.
How To Deliver Your Message On Time

"It's really clear that the most precious resource we all have is time."

- STEVE JOBS

I feel like I'm gonna die...

60 seconds left... I still have two points to make... What should I do?

I take a deep breath... rush through the slides and... I have missed the mark.

The audience hated me for going over time. As a leader, I lost trust and credibility.

Don't want to (feel like you're gonna) die? Follow these three effective methods and gain more trust and credibility.

1. FIVE MINUTES PER POINT

A point supports your message. Most business leaders try to make too many points in too little time. As a result, they don't have enough time to explain all points.

Successful leaders explain each point in five minutes. They get to the point and move on to the next one.

2. THREE MAIN POINTS PER PRESENTATION

Threes are memorable. If you make ten points in a presentation, no one will remember any of them. It's better to make three important points than ten not-so-important points.

What are the three main points that support your message?

3. PRACTICE WITH FEEDBACK

Practice doesn't make perfect; practice with feedback makes great improvement.

Get feedback from a coach that can take your presentation skills to the next level.

With coaching, you will deliver the message more effectively and be on time.

It's all about time. Get to the point, get feedback, and stay with the time you're given.

Part III:

REMEMBER WHAT TO SAY

5.

Three Reasons You Didn't Remember What To Say In Presentation

"Memory… is the diary that we all carry about with us."

- OSCAR WILDE

Imagine you're preparing for a presentation. You've memorized the presentation word-for-word. You have hand-written notes on the lectern.

During presentation, everything goes well.. until someone raises his hand and asks you a question… until someone leaves the room… until someone farts.

Suddenly, you forget what to say…

Your face turns red, you're sweating, and you don't know what to do.

Sound familiar?

Here's why you didn't remember what to say during presentation:

1. MEMORIZE A PRESENTATION

Memorizing a presentation is not effective. When someone interrupts us, we suddenly

forget what to say. Not knowing our script can hurt our credibility and trust with the

audience.

SOLUTION: USE NOTES EFFECTIVELY

It's ok to use notes. Throw away your word-for-word script. Use point form. Make sure

the font size is at least 14.

When you look at the notes, pause... Look up and speak again. People will see you as

a leader who is confident and in-control.

2. MAKE TOO MANY POINTS

No more brain dumps... please! My brain can't handle it. I know you want the audience to take action, and you want to have a positive impact on their lives. But more is not more. Less is more.

If you're lazy, good. Make fewer points.

SOLUTION: MAKE THREE POINTS AT MOST

People remember three the best. Only make points that help deliver your message effectively.

Brainstorm ten points, cross out seven and pick the three most important points.

3. USE TOO MANY FACTS AND DATA

Facts and data suck. Nobody has time for another round of data bomb.

These boring facts and data are so complicated that they're (almost) impossible to remember.

SOLUTION: USE STORIES TO MAKE YOUR POINT

We remember points better with 5 senses and emotions. Stories turns random facts into an emotional roller coaster ride.

When you remember the story, you remember the point.

Use notes, make three points, and tell stories. You'll remember what to say, be more productive, and build more trust with people.

6.
How To Practice Your Presentation The Right Way

"Take advantage of every opportunity to practice your communication skills so that when important occasions arise, you will have the gift, the style, the sharpness, the clarity, and the emotions to affect other people."

- JIM ROHN

When I was young, my teacher taught me, "Practice makes perfect."

I used to have a bad memory. I said, "I worry that my mind will go completely blank, and I'll forget what I'm talking about."

I tried to practice in front of a mirror, but all I saw was an angry, red pimple on my face. I felt even more uncomfortable.

In college, I listened to a presentation from Craig Valentine, 1999 World Champion of Public Speaking. He said, "What if you're practicing the wrong habits again and again? You're doomed to fail!"

Here's my discovery: Practice doesn't make perfect; practice with feedback makes great improvement. When you learn how to practice a presentation the right way, you will remember what to say.

1. RECORD YOURSELF IN VIDEO

The worst time to evaluate your presentation is when you're giving it. I recommend you use a smartphone or webcam to record your rehearsal.

My friend, TJ Walker says, "When you're rehearsing, just focus on giving the speech, and then look at the video." As the author of *TJ Walker's Secret to Foolproof Presentations*, he adds, "You really cannot give a speech and be critiquing it at the same time."

2. GET FEEDBACK

Time to have some popcorn and watch your video. I know it can be painful to watch yourself speaking. But if you can't even look at yourself speaking, how can you expect the audience to enjoy your presentation? Just watch the video, and you'll feel more and more comfortable.

Write down what you like and don't like

about your performance, especially body language and use of voice.

Watch the first three to five minutes of your video to get the big picture quickly. Matthew Kohut, author of *Compelling People,* suggests that you can sometimes watch the middle to make sure your energy level is high. Also, make sure you end strong.

3. IMPROVE YOUR PRESENTATION

Christopher Avery, creator of *The Leadership Gift,* recommends keeping things that work and getting rid of things that don't. Like your high energy? Keep it. Hate your poor eye contact? Focus on improving your eye contact. Just speak to the audience as if you're talking to friends.

PART IV:

DELIVER YOUR MESSAGE CONFIDENTLY

7.
Picture Your Audience In Their Underwear — That's A Stupid Strategy

"The essence of strategy is choosing what not to do."

- MICHAEL PORTER

Have you ever heard people say, "You should picture the audience in their underwear?"

Oh my God! That's so weird. As a professional, stop doing that... please.

Here's a more effective way to overcome your public speaking anxiety, easily and quickly: **Picture that you're talking to your friends in the living room.**

Imagine you're talking to your friends in the living room... they laugh at your jokes and find you interesting.

"Yeah, I think about that too. Wow you're amazing."

How do you feel? You feel less anxious and more confident.

When delivering your message, focus on your audience. Treat them as your friends, just like you're having a conversation with them.

Next time, when delivering your presentation, don't picture your audience in their underwear. Picture them as your friends.

8.
The Powerful Formula To Deliver Your Message With Confidence

"Confidence comes from discipline and training."
- ROBERT KIYOSAKI

Imagine you're delivering the message. You feel confident when presenting your material. People see you as a leader. You inspire people and change their behaviour.

You feel confident, empowered, and in-control.

This doesn't have to be in your dreams. Stop worrying about what people think about you. You can deliver your message confidently with this powerful formula.

THE PERFECT PRESENTATION DAY FORMULA

1. TURN YOUR NERVOUSNESS INTO EXCITEMENT

Christopher Avery, leadership expert, is a pro at reframing bad situations. When facing stress, you have two options: you can either feel scared to death or feel excited about the opportunity to share your message with the world.

I choose to turn my stress into excitement because I want to transform people's lives and help them become more successful. How about you?

2. ARRIVE EARLY AT THE VENUE

Arrive 15 to 20 minutes early to where you'll deliver the presentation. You may want to get comfortable with the stage, especially if you've never been there before.

Test your equipment. Make sure the computer and projector is working. If you are playing video, check the sound. For big audiences, use a microphone.

Practice your opening so you will gain

more confidence and comfort.

Talk to your audience. Instead of presenting in front of a group of strangers, why not build relationships with your audience before speaking? You'll turn them from strangers into friends and feel more relaxed.

3. RELAX YOUR BODY

Go to the toilet and stay away from distractions. Relax your jaw, so it will not feel tense. Relax your head, so you will feel comfortable. Relax your shoulders, hands, knees, and feet. Now you're calm and relaxed.

4. BREATHE IN SLOWLY AND DEEPLY

Taking slow and deep breaths helps you relax. It lowers your stress hormones, drops your heart rate, and makes you feel more comfortable.

Deep breathing also gives you more energy by bringing more oxygen to your blood. Breathe.

5. BE IN THE MOMENT

Focusing on the audience is a proven method to improve confidence. Fred Miller from *No Sweat Public Speaking* shares, "If you focus on the audience and providing those needs for them, the quality of your presentation will go up, and your anxiety will go down."

Before stepping on stage, I focus on helping other people become successful and on making a difference. After all, it's not about you; it's about the audience.

PART V:

INSPIRE PEOPLE TO TAKE ACTION

9.
Why Most People Don't Take Action (And What To Do Instead)

"Every action needs to be prompted by a motive."

- LEONARDO DA VINCI

I hate wasting my time!

I spent weeks preparing for this presentation. They love me... but why didn't they take action and buy my products?

Embarrassed, I asked one of the audience members. He smiled and said, "Because you didn't ask us to."

I smashed my head to the wall... nearly (If I did, you wouldn't be reading this book).

The goal of presentation is to inspire

action.

Only when the audience takes action on what you're selling, will they will improve their lives.

These 3 reasons explain why most people don't take action:

1. WEAK CLOSING

"That's the end of my presentation. Thank you." - is the worst closing ever.

People remember the opening and closing the best. We want to end strong and get the audience to take action.

SOLUTION: HAVE A STRONG CLOSING

To end strong, have a clear summary, memorable message, and call to action. When you include all these 3 parts in the closing, you will inspire action and have the impact you want on the audience.

2. NO CALL TO ACTION

Do you hate the feeling that you want to

buy a product, but don't know what to do?

The audience members are ready to take action, only when you tell them the next step.

Don't leave the audience hanging. Tell them *exactly* what to do.

SOLUTION: HAVE A CLEAR CALL TO ACTION

A call to action can be as simple as "Fill out this application form and give it back to me by the end of this meeting."

Be direct. Speak your words and inspire action.

3. END WITH Q&A

Follow me: the audience remembers your opening and ending the best.

If you end with Q&A, will the audience remember all the points you talked about? I doubt that.

SOLUTION: HAVE Q&A BEFORE CLOSING

Near the end of your presentation, say "Before I close, I'm going to take a few questions. What questions do you have?"

Then move on to the closing. The audience will remember what you say and take action.

10.
Three Essential Steps For Inspiring Your Audience To Take Action

"If your actions inspire others to dream more, learn more, do more and become more, you are a leader."

- JOHN QUINCY ADAMS

I'll never forget the day I gave a 20-minute talk at a local service club, and no one bought from me...

Have you ever heard people say that you should speak for free? I took the advice and was frustrated about being forgettable. Here I was preparing a presentation and getting no customers. "Free presentations are bad for my business. What a waste of time!"

Hearing me complain about my failure, my entrepreneur friend said, "Free presentations are not the problem. The problem is you don't have a memorable closing that inspires people to take action."

But no one had given me a step-by-step for inspiring action. I ended up studying dozens of the most popular TED talks and Apple presentations, and I found out that there's a pattern among all these powerful closings. I modelled their success and managed to inspire my audience to take action. After I spoke, passionate audience members rushed to me, handing me their business cards, begging me to work with them.

Inspire your audience to take action with these three essential steps:

1. SUMMARIZE KEY POINTS

People remember your closing the best. To recall their memory, do a quick recap. If my presentation is about inspiring action, I would summarize my points with, "How can we inspire people to take action? Number one, summarize your key points. Number two, repeat your message. Number three, have a clear call to action."

2. REPEAT YOUR MESSAGE

The audience cannot remember what you said immediately. Repeat your message at least two times: once in the story and once in the closing. In my TEDx talk, I repeat my message in the closing: "How do we speak with confidence? Prepare, practice, perform."

3. HAVE CALL TO ACTION

Learning is meaningless if we don't take action. Scott Schewerly, CEO of a presentation design company named Ethos3, asks, "If you don't have a call to action within your talk, then why in the world do you give it?" No matter how great your presentation is, having no audience member take action is a waste of your time.

Your audience may not know what to do after your presentation. Give them a clear action step. The action step can be as simple as, "Go to XYZ.com and download my free report."

To inspire people to take action, have a summary, message, and call to action. When your audience knows what to do, they will take action.

BIBLIOGRAPHY

Berkun, Scott. *Confessions of A Public Speaker*. California: O'Reilly, 2010. Print.

Burchard, Brendon. *The Millionaire Messenger: Make a Difference and a Fortune Sharing Your Advice.* London: Simon & Schuster, 2011. Print.

Carnegie, Dale. *How to Win Friends & Influence People.* New York: Pocket, 1998. Print.

Carnegie, Dale. *The Quick & Easy Way to Effective Speaking.* New York: Pocket, 1962. Print.

Carter, Judy. *Stand-Up Comedy: The Book.* New York: Delta, 1989. Print.

Carter, Judy. *The Message of You: Turn Your Life Story into a Money-Making Speaking Career*, 2013. Web. 24 Sept 2014.

Donovan, Jeremey. *How to Deliver A TED Talk: Secrets of the World's Most Inspiring Presentations.* Columbus: McGraw-Hill, 2014. Print.

Esposito, Janet. *In The Spotlight: Overcome Your Fear of Public Speaking and Performing*, 2008. Web. 25 Oct. 2013.

Gallo, Carmine. *Talk Like TED: The 9 Public Speaking Secrets of The World's Top Minds.* New York: Macmillan, 2014. Print.

Gallo, Carmine. *The Presentation Secrets of Steve Jobs: How to Be Insanely Great in Front of Any Audience.* New York: McGraw-Hill, 2010. Print.

Karia, Akash. *How to Deliver the Perfect TED Talk: Presentation Secrets of the World's Best Speakers*, 2013. Web. 21 Nov. 2013.

Kawasaki, Guy. *Enchantment: How To Woo, Influence And Persuade.* London: Penguin, 2011. Print.

Marshall, Lisa B. *Smart Talk: How to Make Genuine Conversations, Build Lasting Relationships and Influence Others.* New York, 2013. Griffin, 2013. Web. 4 Jan. 2013.

Maxwell, John C. *Everyone Communicates, Few Connect: What the Most Effective People Do Differently.* Nashville: Thomas Nelson, 2010. Print.

Miller, Fred. *No Sweat Public Speaking*, 2011. Web. 7 Apr. 2013.

Monarth, Harrison, and Larina Kase. *The Confident Speaker: Beat Your Nerves and Communicate at Your Best in Any Situation*. New York: McGraw-Hill, 2007. Print.

Neffinger, John, and Matthew Kohut. *Compelling People: The Hidden Qualities That Make Us Influential*. London: Piatkus, 2013. Print.

Reynolds, Garr. *Presentation Zen: Simple Ideas on Presentation Design and Delivery*. Berkeley: New Riders, 2012. Print.

Sedniev, Andrii. *Magic of Impromptu Speaking: Create a Speech That Will Be Remembered for Years in Under 30 Seconds*, 2014. Web. 25 Sept. 2014.

Simmons, Annette. *The Story Factor: Inspiration, Influence, and Persuasion Through the Art of Storytelling*. New York: Basic Books, 2001. Print.

Tracy, Brian. *Speak to Win: How to Present with Power in Any Situation*. New York: AMACOM, 2008. Print.

Valentine, Craig, and Mitch Meyerson. *World Class Speaking: The Ultimate Guide to Presenting, Marketing and Profiting like a Champion.* New York: Morgan, 2009. Print.

Walker, TJ. *How to Give A Pretty Good Presentation.* New Jersey: Wiley, 2010. Print.

Weissman, Jerry. *Presenting to Win: The Art of Telling Your Story.* New Jersey: Pearson, 2009. Print.

About the Author

Jonathan Li helps business leaders deliver their message effectively and confidently, and have the impact they want on their audience.

He has been featured on TEDxHongKong and Entrepreneur.com

You can reach Jonathan at TheExpressiveLeader.com